9th International Conference on Upstream Engineering and Flow Assurance 2021

Topical Conference at the 2021 AIChE Spring Meeting and 17th Global Congress on Process Safety

Online
18 – 23 April 2021

ISBN: 978-1-7138-5334-3

Printed from e-media with permission by:

Curran Associates, Inc.
57 Morehouse Lane
Red Hook, NY 12571

Some format issues inherent in the e-media version may also appear in this print version.

Printed with permission by Curran Associates, Inc. (2022)

For permission requests, please contact AIChE
at the address below.

AIChE
120 Wall Street, FL 23
New York, NY 10005-4020

Phone: (800) 242-4363
Fax: (203) 775-5177

www.aiche.org

Additional copies of this publication are available from:

Curran Associates, Inc.
57 Morehouse Lane
Red Hook, NY 12571 USA
Phone: 845-758-0400
Fax: 845-758-2633
Email: curran@proceedings.com
Web: www.proceedings.com

TABLE OF CONTENTS

618424 Measurement and Modeling of the Viscosity and Stability of Visbroken Fractionated Bitumen

Abstract:

In Alberta, bitumen production is restricted by the limited pipeline capacity to external markets. The capacity is further limited because the bitumen must be diluted with a solvent to meet the pipelines' viscosity and density specifications [1]. Solvents occupy 25% to 50% of the bitumen transportation pipelines and require a second pipeline to return the solvents [2]. One way to increase the capacity is to use a combination of partial deasphalting and visbreaking [3,4] to reduce the density and viscosity of the bitumen so that less diluent is required. Partial deasphalting uses a poor solvent to precipitate and remove some of the asphaltenes from the bitumen. The asphaltenes are the densest fraction of the bitumen; hence deasphalting reduces the product density but only moderately reduces its viscosity. Vibreaking is a mild liquid-phase thermal cracking method that only moderately reduces the density the bitumen but significantly reduces its viscosity [5,6]. The viscosity reduction that can be achieved in a visbreaking process depends on the intensity (residence time and temperature) of the process. There is an optimum intensity below which there is less viscosity reduction and above which coke forms. Coke formation is undesirable because it leads to fouling. The visbroken product must also be stable versus asphaltene precipitation when blended with a diluent. Therefore, to optimize a visbreaking process it is necessary to predict the viscosity and stability (versus asphaltene precipitation) of the product.

The eventual goal of this project is to examine combinations of visbreaking and deasphalting. As a first step, fractionated oils are examined and the specific objective of this study is to measure and model the viscosity and stability of the products from three types of visbreaking process feeds: a whole oil, a vacuum residue, and a partially deasphalted oil. The feeds are all from or derived from Western Canadian bitumen. A methodology designed for integration with process simulation software and previously applied only to whole bitumen was followed [6,7]. Each of the feeds were reacted in an in-house visbreaker at two different conversions. The composition of the products were measured in terms of distillates and the SARA fractions of the distillation residue. The molecular weight, density, and viscosity of each fraction from the feeds were measured. The density, viscosity, and asphaltene solubility of the products and there maltene and residue fractions were also measured. The Expanded Fluid viscosity model was used to model the viscosity of the products. The Modified Regular Solution model was used to model stability of the products.

The Expanded Fluid model relates viscosity to density using three fluid specific parameters. The model includes mixing rules to determine whole oil parameters from the oil composition and component parameters. In a previous study [6], recommended parameters were proposed for the distillates and SARA fractions from whole bitumen. The density of each fraction (a required input) was related to temperature using two parameters and recommend values were provided. In addition, correlations were developed to predict the viscosity and density parameters as a function of conversion for fractions from visbroken oils. Preliminary results indicate that the recommended parameters and correlations also apply to vacuum bottoms and deasphalted oils except for the asphaltenes from deasphalted oils. Notable property trends, the differences from the previous properties, and the proposed modifications to the correlations will be discussed.

The Modified Regular Solution model is an activity-coefficient-based liquid-liquid phase equilibrium model which requires the molecular weight, density, and solubility parameter of each component in the mixture. In a previous study [7], recommended properties were proposed for the distillates and SARA fractions from whole bitumen. In addition, correlations were developed to predict these properties as a function of conversion for fractions from visbroken oils. Preliminary results indicate that the recommended properties and correlations also apply to vacuum bottoms and deasphalted oils except for the asphaltenes from deasphalted oils. Notable property trends, the differences from the previous properties, and the proposed modifications to the correlations will be discussed.

[1] R. W. Luhning, A. Anand, T. Blackmore, and D. S. Lawson, "Pipeline transportation of emerging partially upgraded bitumen," in *Canadian International Petroleum Conference 2002, CIPC 2002*, 2002.

[2] A. Hart, "A review of technologies for transporting heavy crude oil and bitumen via pipelines," *Journal of Petroleum Exploration and Production Technology*. 2014.

[3] M. R. Gray, *Upgrading Oilsands Bitumen and Heavy Oil*. The University of Alberta Press, 2015.

[4] A. Zachariah and A. De Klerk, "Partial Upgrading of Bitumen: Impact of Solvent Deasphalting and Visbreaking Sequence," *Energy and Fuels*, 2017.

[5] J. B. Joshi *et al.*, "Petroleum residue upgradation via visbreaking: A review," *Industrial and Engineering Chemistry Research*. 2008.

[6] A. Marquez, F. F. Schoeggl, S. D. Taylor, G. Hay, and H. W. Yarranton, "Viscosity of characterized visbroken heavy oils," *Fuel*, 2020.

[7] S. Rodriguez, E. N. Baydak, F. F. Schoeggl, S. D. Taylor, G. Hay, and H. W. Yarranton, "Regular solution based approach to modeling asphaltene precipitation from native and reacted oils: Part 3, visbroken oils," *Fuel*, 2019.

618889 The Effect of Methane on Asphaltene Precipitation

Abstract:

Asphaltenes are the heaviest, most polar, and most aromatic fraction of crude oil and can form a separate phase from the crude oil due to changes in pressure, temperature, or oil composition [1]. Asphaltene precipitation is usually undesirable because it can lead to formation damage and fouling of pipelines and production facilities. However, it is a key step in some processes such as deasphalting. In either case, it is necessary to identify the conditions at which asphaltenes begin to precipitate and in some cases the amount of precipitation. Two well-known situations where asphaltene precipitation can occur during the production and processing of crude oils are: 1) diluting a heavy oil with an incompatible solvent (*e.g.* an *n*-alkane); 2) depressurizing a light conventional oil. *In situ* heavy and live conventional oils both contain dissolved gases, particularly methane, which have a significant negative impact on the oil's ability to solubilize asphaltenes [2, 3].

One approach to modeling asphaltene precipitation is the modified regular solution (MRS) liquid-liquid equilibrium model. The MRS model has been used previously to model asphaltene precipitation from heavy oils and bitumen diluted with *n*-alkanes at different temperature and pressures [4]. The oils are characterized into SARA fractions (saturates, aromatics, resins, and pentane insoluble asphaltenes which are termed C_5-asphaltenes). The inputs parameters for the model are the mole fraction, molar volume and solubility parameter for each SARA fraction and each solvent. The current version of the regular solution model is not yet able to predict asphaltene solubility in the presence of dissolved gases because it lacks accurate solubility parameters of these dissolved gases. The objective of this project is to quantify and model the effect of methane on asphaltene solubility and, in particular, to determine the solubility parameter of the dissolved methane.

The solubility parameters for a solvent can be determined by fitting a Modified Regular Solution model (MRS) to the asphaltene yield curves from heavy oil diluted with the solvent of interest. A yield curve is a plot of the yield (mass of precipitated divided by mass of feed oil) versus the mass fraction of solvent in the mixture. However, the methane solubility in a bitumen is too low to trigger asphaltene precipitation. Instead, the methane solubility parameter was determined from asphaltene yield curves for mixtures of bitumen, *n*-pentane, and methane. The properties of the bitumen and *n*-pentane required for the MRS model have already been established [5]. The density of the dissolved methane was determined using an effective density correlation. Therefore, the only unknown property was the solubility parameter the dissolved methane.

The solvent content at which a heavy asphaltene-rich liquid phase formed (onset) and the asphaltene precipitation yield were determined for mixtures of bitumen, *n*-pentane, and methane at temperatures from 20 to 130 °C and pressures from 10 to 60 MPa. The onsets were visually detected in a High-Pressure Microscope [6] from titrations of bitumen with a mixture of methane and *n*-pentane. Asphaltene yields were measured in a blind cell PVT apparatus following established procedures [7].

The measured onsets were lower than onsets in pure *n*-pentane at similar conditions, confirming that asphaltenes are less soluble in methane than in *n*-pentane. The C_5-asphaltene yields were the same at all conditions, suggesting that all of the C_5-asphaltenes partitioned into the heavy phase at all of the experimental conditions. The data collected to date were modeled and fitted with a methane solubility parameter of approximately 8.5 $MPa^{0.5}$. This value is much lower than solubility parameter of *n*-pentane (14.4 $MPa^{0.5}$) and is consistent with the significantly poorer solubility of asphaltenes in methane. The effect of temperature and pressure on the methane solubility parameter will be discussed. The impact of methane content on the onset of asphaltene precipitation will be illustrated.

References

[1] A. K. Tharanivasan, H. W. Yarranton, and S. D. Taylor, "Application of a regular solution-based model to asphaltene precipitation from live oils," *Energy and Fuels*, vol. 25, no. 2, pp. 528–538, 2011.

[2] S. A. M. Dehghani, M. V. Sefti, and G. A. Mansoori, "Simulation of natural depletion and miscible gas injection effects on asphaltene stability in petroleum reservoir fluids," *Pet. Sci. Technol.*, vol. 25, no. 11, pp. 1435–1446, 2007.

[3] P. Zanganeh, H. Dashti, and S. Ayatollahi, "Comparing the effects of CH4, CO2, and N2 injection on asphaltene precipitation and deposition at reservoir condition: A visual and modeling study," *Fuel*, vol. 217, no. January, pp. 633–641, 2018.

[4] K. Akbarzadeh, H. Alboudwarej, W. Y. Svrcek, and H. W. Yarranton, "A generalized regular solution model for asphaltene precipitation from n-alkane diluted heavy oils and bitumens," *Fluid Phase Equilib.*, vol. 232, no. 1–2, pp. 159–170, 2005.

[5] F. Ramos-Pallares and H. W. Yarranton, "Extending the modified regular solution model to predict component partitioning to the asphaltene-rich phase," *Energy and Fuels*, 2020.

[6] P. Agrawal, F. F. Schoeggl, M. A. Satyro, S. D. Taylor, and H. W. Yarranton, "Measurement and modeling of the phase behavior of solvent diluted bitumens," *Fluid Phase Equilib.*, vol. 334, pp. 51–64, 2012.

[7] K. A. Johnston, F. F. Schoeggl, M. A. Satyro, S. D. Taylor, and H. W. Yarranton, "Phase behavior of bitumen and n-pentane," *Fluid Phase Equilib.*, vol. 442, pp. 1–19, 2017.

619302 Stability of Asphaltenes in Oils and Fuels

Abstract:

Effective from 1st January 2020, the International Maritime Organization (IMO) reduced the maximum allowable Sulphur content in marine fuel oil from 3.5 % to 0.5% by mass. One approach that can be taken to comply with this regulation is blending feedstocks to obtain a fuel blend with the required Sulphur content. However, blending various feedstocks can result in a mixture that contains unstable asphaltenes. In turn, unstable asphaltenes can foul equipment and will ultimately impair engine and machinery performance. In this work, standardized tests used in industry to evaluate the stability of asphaltenes in oils and fuels will be discussed. In addition, the recent findings on asphaltene aggregation and fouling will be presented and their impact on the standardized tests will be discussed.

618728 Investigating Asphaltene Deposition in Pipe Flow at Elevated Temperatures

Abstract:

Asphaltene deposition is a longstanding flow assurance issue. It has been studied extensively at near ambient temperatures where asphaltenes are commonly considered to precipitate as glassy particles [1-7]. Asphaltene deposition at these conditions involves the adhesion and accumulation of asphaltene molecules, particles, and aggregates at the surface of a pipe or vessel. Several models have been developed to capture asphaltene deposition mechanisms accounting for asphaltene aggregate size and concentration, shear conditions, fluid properties, and surface properties [8-9]. However, asphaltene deposition at higher temperatures encountered in deep, offshore production has not been rigorously investigated. Precipitated asphaltenes at higher temperatures are liquids rather than glassy particles and may deposit differently. The goals of this project are to: 1) design and commission a deposition test apparatus for temperatures up to 130°C; 2) measure the effects of asphaltene solubility, asphaltene-rich phase properties, time, and adhesion on deposition rates; develop a model or correlation relating deposition rates to changes in temperature, pressure, flow rates, and solvent composition. In this contribution, the following are presented: design of the proposed deposition apparatus, validation with deposition measurements at 50°C, and preliminary results at higher temperatures. To date, deposition experiments have been performed in a horizontal flow configuration with a Western Canadian bitumen mixed with *n*-heptane.

The new apparatus was designed to measure asphaltene deposition from once-through flow of mixtures of oil and solvent at temperatures from 50 to 130°C and pressures up to 10 MPa. Solvent induced precipitation was selected as a design basis over pressure drop induced precipitation to obtain measurable and easily controlled quantities of precipitate. While the apparatus does not directly match offshore wellbore conditions, it permits an examination of deposition mechanisms at elevated temperatures. Deposition is detected via a change in the pressure drop through a 30 cm long, 1.4 mm I.D. tube (the test section). The mass and location of the deposit can be determined by removing the test section. The diameter of the tube was small enough to observe a measurable incremental pressure drop with deposition and large enough to obtain a quantifiable mass measurement.

The apparatus consists of two transfer vessels, a custom built co-axial nozzle-like static mixer, and the test section. A Quizix SP-52000 pump system and a Teledyne ISCO Demi 2510S pump are used to displace pre-diluted bitumen and *n*-heptane, respectively, from the transfer vessels to the test section. The differential pressure between the inlet and outlet of the capillary tube is measured with a Rosemount 1151 pressure transducer. The apparatus is housed inside an air bath.

In each deposition experiment, the bitumen was pre-diluted with *n*-heptane to a point below the onset of precipitation to reduce its viscosity and promote rapid mixing. Then, the pre-diluted bitumen and additional solvent was fed through the static mixer and displaced through the capillary tube. The pressure drop across the tube was monitored during the flow period. At the end of the flow period, the free liquid in the tube was displaced with cyclohexane, depressurized and disconnected, and the mass of deposit determined from the change in mass of the tube. In some experiments, the tube was cut open to examine the location of the deposit. The controlled variables are the type of solvent, solvent/oil ratio, temperature (operating range of 20 to 130°C), pressure (up to 10 MPa), and flow rate (2 to 5 cm^3/min).

A critical factor in these experiments was the mixing of the bitumen and solvent. The effectiveness of the static mixer was assessed by comparing asphaltene yields (mass of precipitated divided by mass of bitumen in feed) at the exit of the mixer with asphaltene yields measured in test tubes all at ambient conditions. Poor mixing in the static mixer would give lower yields than observed in the well mixed test tube tests. The yields at the exit of the mixer matched the test tube yields confirming effective mixing above the onset of precipitation.

However, an issue with mixing at solvent/bitumen ratios below the onset of precipitation was identified. As expected, no precipitation occurred at these conditions in the test tube tests. Similarly, any pressure drop observed in the apparatus for a feed of premixed bitumen/solvent below the onset ratio was less than the error of the measurement. However, pressure drops indicative of deposition were observed in the deposition apparatus when bitumen and solvent were mixed in the static mixer even when the final composition was below the onset condition. A potential explanation is that there are local high solvent content zones within the mixer that cause local precipitation and subsequent deposition. Therefore, any data collected below the onset of precipitation are suspect.

The data collected above the onset of precipitation at 50°C were consistent with the literature. The pressure drop measurements indicated cycles of deposition and erosion during the flow period. Asphaltene deposits were found to be concentrated near the inlet of the capillary tube. Preliminary data collected at higher temperatures will be presented and the effect of higher temperatures on the deposition rates will be discussed.

References

1. Broseta, D., Robin, M., Savvidis, T., Féjean, C., Durandeau, M., Zhou, H. Detection of Asphaltene Deposition by Capillary Flow Measurements. SPE paper 59294, SPE/DOE Improved Oil Recovery Symposium, Tulsa, Oklahoma, 3–5 April, 2000.

2. Wang, J., Buckley, J.S., Creek, J.L. Asphaltene Deposition on Metallic Surfaces. *Dispersion Sci. Techno.*, 25:3, 2004, 287-298.

3. Nabzar, L., Aguilera, M.E. The Colloidal Approach. A Promising Route for Asphaltene Deposition Modelling. *Oil & Gas Science and Technology – Rev. IFP*, 63, 2008, 21-35.

4. Seifried, C.M., Al Lawati, S., Crawshaw, J.P., Boek, E.S. Asphaltene Deposition in Capillary Flow. SPE paper 166289, SPE Annual Technical Conference and Exhibition, New Orleans, Louisiana, USA, 30 September – 2 October, 2013.

5. Hoepfner, M.P., Limsakoune, V., Chuenmeechao, V., Maqbool, T., Fogler, H.S. A Fundamental Study of Asphaltene Deposition. *Energy Fuels*, 27, 2013, 725-735.

6. Bo̕as Fa̕vero, C.V., Hanpan, A. Phichphimok, P., Binabdullah, K., Fogler, H.S. Mechanistic Investigation of Asphaltene Deposition. *Energy Fuels*, 30, 2016, 8915-8921.

7. Ghahfarokhi, A.K., Kor, P., Kharrat, R., Soulgani, B.S. Characterization of asphaltene deposition process in flow loop apparatus; An experimental investigation and modeling approach. *Petr. Sci. Eng.* 151, 2017, 330-340.

8. Vargas, F.M., Creek, J.L., Chapman, W.G. On the Development of an Asphaltene Deposition Simulator. *Energy Fuels*, 24, 2010, 2294-2299.

9. Eskin, D., Ratulowski, J., Akbarzadeh, K., Pan, S. Modelling Asphaltene Deposition in Turbulent

Pipeline Flows. *J. Chem. Eng.*, 89, 2011, 421-441.

NOTES

NOTES

AUTHOR INDEX